YOUR KNOWLEDGE HAS VALUE

General Permutation Theorem and Selected Permutations

Deapon Biswas

Bibliographic information published by the German National Library:

The German National Library lists this publication in the National Bibliography; detailed bibliographic data are available on the Internet at http://dnb.dnb.de.

ISBN: 9783389011782
This book is also available as an ebook.

© GRIN Publishing GmbH
Trappentreustraße 1
80339 München

Print and binding: Books on Demand GmbH, Norderstedt, Germany
Printed on acid-free paper from responsible sources.

The present work has been carefully prepared. Nevertheless, authors and publishers do not incur liability for the correctness of information, notes, links and advice as well as any printing errors.

GRIN web shop: https://www.grin.com/document/1464424

General Permutation Theorem and Selected Permutations

Deapon Biswas

Transport Officer, Private Concern, Chattogram, Bangladesh

Abstract

We have permutations are discussed with different theorems in algebra. In this chapter I apply B system analysis to get the theorems easy and memorable. After B system analysis applied there becomes a lot of new theorems and all the theorems get a new face by summation methods. One usual theorem described with summation method and face new looks.

Keywords

Permutation space, permutation member, permutation event, permutation partial space, permutation partial event type I, permutation partial event type II, general permutation theorem, permutation distribution, selected permutations.

Article Outline

1. Introduction
2. Preliminaries
3. Permutation Space
4. Permutation Member
5. Permutation Theorem
6. Permutation Event
7. Identified Permutation Theorem
8. Permutation Partial Space
9. Permutation Partial Event Type I
10. Permutation Partial Event Type II
11. General Permutation Theorem
12. Permutation Distribution
13. Selected Permutations
14. Selected Permutation Theorem
15. Identified Selected Permutation Theorem
16. Conclusion

1. Introduction

We have a full idea about permutation. It indicates the outcome of a random experiment. That is permutation is the selection of M different components taken V at a time what is called usually a random experiment where order is taken into account and repetitions are not allowed.

2. Preliminaries

As a preliminary it may be stated the following theorem

$$P\binom{N}{V} = N(N-1)(N-2) \dots (N-V+1); \quad V \le N \quad\quad\quad (1)$$

3. Permutation. Space

A permutation space is a set of all possible permutations (outcomes) of an experiment from a parent assembly A where the outcomes take order of the components into account. Let a permutation space contains T possible outcomes then the permutation space denoted by $P\left\{\begin{matrix}A\\V\end{matrix}\right\}$ is

$$P\left\{\begin{matrix}A\\V\end{matrix}\right\} = \{P_1, P_2, P_3, \dots, P_t, \dots, P_T\} \quad\quad\quad\quad (2)$$

where, V is the number of components occurred in an outcome.

Example 1: Set a permutation space of the experiment "arrange 4 digits 1, 2, 3 and 4 taken 3 at a time".

Solution: We have given A = (1, 2, 3, 4) and V = 3.

Thus the permutation space is

$$P\left\{\begin{matrix}(1,2,3,4)\\3\end{matrix}\right\} = \{(1, 2, 3), (1, 3, 2), (1, 2, 4), (1, 4, 2), (1, 3, 4), (1, 4, 3),$$

(2, 3, 4), (2, 4, 3), (2, 3, 1), (2, 1, 3), (2, 4, 1), (2, 1, 4), (3, 4, 1), (3, 1, 4), (3, 4, 2), (3, 2, 4), (3, 1, 2), (3, 2, 1), (4, 1, 2), (4, 2, 1), (4, 1, 3), (4, 3, 1), (4, 2, 3), (4, 3, 2)}.

4. Permutation Member

A permutation member is an element of the permutation space (2) usually denoted by P_t ; t = 1, 2, 3 ,……., T ; is

$$P_t = (P_{t1}, P_{t2}, P_{t3}, \dots, P_{tv}, \dots, P_{tV}) \quad\quad\quad\quad (3)$$

The permutation member (16.4.2) contains V permutation components. For everyday use we use the word "permutation" to mean permutation member.

Example 2: Find the permutations P_1, P_9, P_{14} and P_{20} of the example 1.

Solution: The permutations are $P_1 = (1, 2, 3)$, $P_9 = (2, 3, 1)$, $P_{14} = (3, 1, 4)$ and $P_{20} = (4, 2, 1)$.

5. Permutation Theorem

Theorem 1: The number of permutations of N different components taken V at a time denoted by $P\binom{N}{V}$ is

$$P\binom{N}{V} = N(N-1)(N-2)... (N-V+1) \quad\qquad (4)$$

Using the summation method we get (4) as

$$P\binom{N}{V} = \sum_{k_1=1}^{N} \sum_{k_2=1}^{(N-1)} \sum_{k_3=1}^{(N-2)} .....\sum_{k_V=1}^{(N-V+1)} C \quad\qquad (5)$$

Proof: Suppose we have N distinct components and V places in a permutation to fill. We arranged the N components in a O. P. A. as first component, second component, third component and so on N^{th} component. The row can be designed as

$$1^{st} \quad 2^{nd} \quad 3^{rd} \qquad n^{th} \qquad N^{th}$$

Fig. 1

And we shall call it ordered O. P. A. We fill the first place with one of the N components of the O. P. A. that begins from first component and ends in N^{th} component. Now k_1 states an index that indicates a component of the O. P. A. taken by first place of the permutation holds the interval

$$1 \leq k_1 \leq N \quad\qquad (6)$$

The term limits introduced here to express an interval taking integral values from lower limit to upper limit. Thus the number of events of permutations characterizing first components denoted by S_1 is

$$S_1 = N - 1 + 1 = N \quad\qquad (7)$$

Using the summation method we get (7) as

$$S_1 = \sum_{k_1=1}^{N} C = N.C = N \quad\qquad (8)$$

C takes unit value.

After filling first place with $k_1{}^{th}$ component the second place can be filled with one of the $\{(k_1 - 1) - 1 + 1 + N - (k_1 + 1) + 1\}$ components of the O. P. A. that begins from first component and ends in $(k_1 - 1)^{th}$ component or (inclusive sence) begins from $(k_1 + 1)^{th}$ component and ends in N^{th} component where $1 \le k_1 \le N$. Now k_2 states an index that indicates a component of the O. P. A. taken by second place of the permutation, holds the limits

$$1 \le k_2 \le k_1 - 1 \qquad\qquad\qquad \text{————————} \quad (9)$$
$$k_1 + 1 \le k_2 \le N$$

where, $1 \le k_1 \le N$. If the condition changes then the intervals will be changed.

first place takes the component

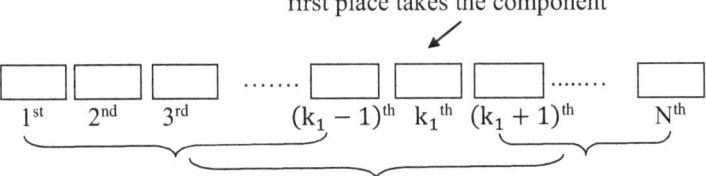

second place takes the component

Fig. 2

Clearly when,

$k_1 = 1$ then ———— $2 \le k_2 \le N$; N–1

$k_1 = 2$ then $1 \le k_2 \le 1$ or, $3 \le k_2 \le N$; 1 + N–2

$k_1 = 3$ then $1 \le k_2 \le 2$ or, $4 \le k_2 \le N$; 2 + N–3

⋮

$k_1 = N{-}1$ then $1 \le k_2 \le N{-}2$ or, $N \le k_2 \le N$; N–2 + 1

$k_1 = N$ then $1 \le k_2 \le N{-}1$ ———— ; N–1

The number after semicolons give the number of indexes held by the corresponding intervals. Thus the number of events of permutations characterizing second components denoted by S_2 is the sum of numbers of indexes held by the intervals (9) i.e.,

$$S_2 = (N{-}1) + (1 + N{-}2) + (2 + N{-}3) + ... + (N{-}2 + 1) + (N{-}1)$$
$$= (N{-}1) + (N{-}1) + (N{-}1) + ... + (N{-}1) + (N{-}1) \qquad \text{(N terms)}$$
$$= N(N{-}1) \qquad\qquad \text{————————} \quad (10)$$

Using the Summation method we get (10) as

$$S_2 = \sum_{k_1=1}^{N} \sum_{k_2=1}^{N-1} C \hspace{3cm} (11)$$

Again filling first place and second place with k_1^{th} and k_2^{th} components respectively we can fill the third place with one of the $\{(k_1 - 1) - 1 + 1 + (k_2 - 1) - (k_1 + 1) + 1 + N - (k_2 + 1) + 1\}$ components of the O. P. A. that begins from first component and ends in $(k_1 - 1)^{th}$ component or begins from $(k_1 + 1)^{th}$ component and ends in $(k_2 - 1))^{th}$ component or begins from $(k_2 + 1)^{th}$ component and ends in N^{th} component where, $1 \le k_1 \le k_2 \le N$. Now k_3 states an index that indicates a component of the O. P. A. taken by third place of the permutation holds the intervals

$1 \le k_3 \le k_1 - 1$

$k_1 + 1 \le k_3 \le k_2 - 1 \hspace{2.5cm} (12)$

$k_2 + 1 \le k_3 \le N$

where, $1 \le k_1 \le k_2 - N$. If the condition changes then the intervals will be changed.

Clearly when,

$k_1 = 1$ and $k_2 = 2$ then $\hspace{1cm}$ $3 \le k_3 \le N$ $\hspace{1cm}$; N–2

$k_1 = 1$ and $k_2 = 3$ then $2 \le k_3 \le 2$ or, $4 \le k_3 \le N$; $1 + N - 3$

$k_1 = 1$ and $k_2 = 4$ then $2 \le k_3 \le 3$ or, $5 \le k_3 \le N$; $2 + N - 4$

$\vdots$

$k_1 = 1$ and $k_2 = N - 1$ then $2 \le k_3 \le N - 2$ or, $N \le k_3 \le N$; $N - 3 + 1$

$k_1 = 1$ and $k_2 = N$ then $2 \le k_3 \le N - 1$ $\hspace{1cm}$; N–1

when,

$k_1 = 2$ and $k_2 = 1$ then $\hspace{2cm}$ $3 \le k_3 \le N$

$\hspace{6cm}$; N–2

$k_1 = 2$ and $k_2 = 3$ then $1 \le k_3 \le 1$ or, $\hspace{1cm}$ or, $4 \le k_3 \le N$

$\hspace{6cm}$; $1 + N - 3$

$k_1 = 2$ and $k_2 = 4$ then $1 \le k_3 \le 1$ or, $3 \le k_3 \le 3$ or, $5 \le k_3 \le N$

$\hspace{6cm}$; $1 + 1 + N - 4$

$k_1 = 2$ and $k_2 = 5$ then $1 \le k_3 \le 1$ or, $3 \le k_3 \le 4$ or, $6 \le k_3 \le N$

$\hspace{6cm}$; $1 + 2 + N - 5$

$\vdots$

$k_1 = 2$ and $k_2 = N - 1$ then $1 \le k_3 \le 1$ or, $3 \le k_3 \le N - 2$ or, $N \le k_3 \le N$

$\hspace{6cm}$; $1 + N - 4 + 1$

$k_1 = 2$ and $k_2 = N$ then $1 \le k_3 \le 1$ or, $3 \le k_3 \le N - 1$ $\hspace{1cm}$

$$; 1+ N-3$$

when,

$k_1 = 3$ and $k_2 = 1$ then ——— $2 \le k_3 \le 2$ or, $4 \le k_3 \le N$

$$; 1+ N-3$$

$k_1 = 3$ and $k_2 = 2$ then $1 \le k_3 \le 1$ or, ——— or, $4 \le k_3 \le N$

$$; 1 + N-3$$

$k_1 = 3$ and $k_2 = 4$ then $1 \le k_3 \le 2$ or, ——— or, $5 \le k_3 \le N$

$$; 2 + N-4$$

$k_1 = 3$ and $k_2 = 5$ then $1 \le k_3 \le 2$ or, $4 \le k_3 \le 4$ or, $6 \le k_3 \le N$

$$; 2+1+ N-5$$

$k_1 = 3$ and $k_2 = 6$ then $1 \le k_3 \le 2$ or, $4 \le k_3 \le 5$ or, $7 \le k_3 \le N$

$$; 2 + 2 + N-6$$

$\vdots$

$k_1 = 3$ and $k_2 = N-1$ then $1 \le k_3 \le 2$ or, $4 \le k_3 \le N-2$ or, $N \le k_3 \le N$

$$; 2+ N-5 +1$$

$k_1 = 3$ and $k_2 = N$ then $1 \le k_3 \le 2$ or, $4 \le k_3 \le N-1$ or, ———

$$; 2 + N-4$$

Similarly when,

$k_1 = N$ and $k_2 = 1$ then ——— $2 \le k_3 \le N-1$ $; N-2$

$k_1 = N$ and $k_2 = 2$ then $1 \le k_3 \le 1$ or, $3 \le k_3 \le N-1$ $; 1 + N-3$

$k_1 = N$ and $k_2 = 3$ then $1 \le k_3 \le 2$ or, $4 \le k_3 \le N-1$ $; 2 + N-4$

$k_1 = N$ and $k_2 = 4$ then $1 \le k_3 \le 3$ or, $5 \le k_3 \le N-1$ $; 3 + N-5$

$\vdots$

$k_1 = N$ and $k_2 = N-2$ then $1 \le k_3 \le N-3$ or, $N-1 \le k_3 \le N-1$; $N-3 + 1$

$k_1 = N$ and $k_2 = N-1$ then $1 \le k_3 \le N-2$ or, ——— $; N-2$

The numbers after semicolons give the numbers of indexes held by the corresponding intervals. Thus the number of events characterizing third components denoted by S_3 is the sum of numbers of indexes held by the intervals (12) i.e.,

$S_3 = (N-2) + (1+N-3) + (2+N-4) + ... + (N-3+1) + (N-2)$

$+(N-2) +(1+N-3) +(1+1+N-4) +(1+2+N-5) + ... +(1+N-4+1) +(1+N-3)$

$+(1+N-3) +(1+N-3) +(2+N-4) +(2+1+N-5) +(2+2+N-6) +... +(2+N-5+1)$

$+(2+N-4)$

$\vdots$

$+ (N-2) + (1+N-3) + (2+N-4) + (3+N-5) + ... + (N-3+1) + (N-2)$

$$= (N-2) + (N-2) + (N-2) + ... + (N-2) \qquad [(N-1)\text{terms}]$$
$$+ (N-2) + (N-2) + (N-2) + ... + (N-2) \qquad [(N-1)\text{terms}]$$
$$+ (N-2) + (N-2) + (N-2) + ... + (N-2) \qquad [(N-1)\text{terms}]$$
$$\vdots$$
$$+ (N-2) + (N-2) + (N-2) + ... + (N-2) \qquad [(N-1)\text{terms}]$$
$$= (N-1)(N-2) + (N-1)(N-2) + (N-1)(N-2) + ... + (N-1)(N-2) \quad [N \text{ terms}]$$
$$= N(N-1)(N-2) \qquad\qquad\qquad (13)$$

Using the summation method we get (13) as

$$S_3 = \sum_{k_1=1}^{N} \sum_{k_2=1}^{(N-1)} \sum_{k_3=1}^{(N-2)} C \qquad\qquad\qquad (14)$$

Continuing this process we get the number of events of permutations characterizing V^{th} components denoted by S_V is

$$S_V = N(N-1)(N-2)... (N-V+1) \qquad\qquad\qquad (15)$$

Using the summation method we get (15) as

$$S_V = \sum_{k_1=1}^{N} \sum_{k_2=1}^{(N-1)} \sum_{k_3=1}^{(N-2)} ... \cdots \sum_{k_V=1}^{(N-V+1)} C \qquad\qquad\qquad (16)$$

As the last events are of one permutation, so the equations (15) and (16) give the number of permutations of N different components taken V at a time.

Example 3: Find the number of permutations of 5 different components taken 3 at a time.

Solution: We have given $N = 5$ and $V = 3$. So, $N-V + 1 = 5-3 + 1 = 3$. Now from the equation (5) we get

$$P\binom{5}{3} = \sum_{k_1=1}^{5} \sum_{k_2=1}^{4} \sum_{k_3=1}^{3} C$$
$$= \sum_{k_2=1}^{4} \sum_{k_3=1}^{3} C + \sum_{k_2=1}^{4} \sum_{k_3=1}^{3} C + \sum_{k_2=1}^{4} \sum_{k_3=1}^{3} C$$
$$+ \sum_{k_2=1}^{4} \sum_{k_3=1}^{3} C + \sum_{k_2=1}^{4} \sum_{k_3=1}^{3} C$$
$$= \left(\sum_{k_3=1}^{3} C + \sum_{k_3=1}^{3} C + \sum_{k_3=1}^{3} C + \sum_{k_3=1}^{3} C \right)$$
$$+ \left(\sum_{k_3=1}^{3} C + \sum_{k_3=1}^{3} C + \sum_{k_3=1}^{3} C + \sum_{k_3=1}^{3} C \right)$$
$$+ \left(\sum_{k_3=1}^{3} C + \sum_{k_3=1}^{3} C + \sum_{k_3=1}^{3} C + \sum_{k_3=1}^{3} C \right)$$
$$+ \left(\sum_{k_3=1}^{3} C + \sum_{k_3=1}^{3} C + \sum_{k_3=1}^{3} C + \sum_{k_3=1}^{3} C \right)$$
$$+ \left(\sum_{k_3=1}^{3} C + \sum_{k_3=1}^{3} C + \sum_{k_3=1}^{3} C + \sum_{k_3=1}^{3} C \right)$$
$$=(3C+3C+3C+3C) + (3C+3C+3C+3C) + (3C+3C+3C+3C)$$
$$+ (3C+3C+3C+3C) + (3C+3C+3C+3)$$

$$= 12C + 12C + 12C + 12C + 12C$$
$$= 60C$$
$$= 60 \times 1 \quad [\text{taking } C = 1]$$
$$= 60.$$

Theorem 2: The total number of permutations of N different components denoted by $P\binom{N}{V}_{V \in \mathcal{V}}$ is

$$P\binom{N}{V}_{V \in \mathcal{V}} = \sum_V N(N-1)(N-2) \dots (N-V+1) \quad\text{———}\quad (17)$$

$$\text{where } V = 1, 2, 3, \dots, N$$

6. Permutation Event

It is a special kind of subset of a permutation space where the permutation members are to be have same first components, same second components, same third components and so on same v^{th} components. Suppose the permutation members are

$$P_1 = (P_{11}, P_{12}, P_{13}, \dots, P_{1v}, \dots, P_{1V})$$
$$P_2 = (P_{21}, P_{22}, P_{23}, \dots, P_{2v}, \dots, P_{2V})$$
$$P_3 = (P_{31}, P_{32}, P_{33}, \dots, P_{3v}, \dots, P_{3V})$$
$$\vdots$$
$$P_t = (P_{t1}, P_{t2}, P_{t3}, \dots, P_{tv}, \dots, P_{tV})$$
$$\text{where, } P_{11} = P_{21} = P_{31} = \dots = P_{t1}$$
$$P_{12} = P_{22} = P_{32} = \dots = P_{t2}$$
$$P_{13} = P_{23} = P_{33} = \dots = P_{t3}$$
$$\vdots$$
$$P_{1v} = P_{2v} = P_{3v} = \dots = P_{tv}$$

Then the permutation event denoted by $P\left\{\begin{smallmatrix} A \\ V/^vA \end{smallmatrix}\right\}$ consists of $P_1, P_2, P_3, \dots,$ P_t i.e.,

$$P\left\{\begin{matrix} A \\ V/^vA \end{matrix}\right\} = \{P_1, P_2, P_3, \dots, P_t\} \quad\text{———}\quad (18)$$

Here A is a parent assembly containing N components, V is the number of components occurred in a permutation member and ^vA is an identified component assembly containing v identified components.

Example 4: Find the permutation events of the example 1 where the identified components are (i) $^1A = (1)$, (ii) $^1A = (2)$, and $^2A = (3, 1)$.

Solution: The permutation events are

(i) $P\left\{\begin{matrix}(1,2,3)\\3/(1)\end{matrix}\right\} = \{(1, 2, 3), (1, 3, 2)\}$

(ii) $P\left\{\begin{matrix}(1,2,3)\\3/(2)\end{matrix}\right\} = \{(2, 1, 3), (2, 3, 1)\}$

(iii) $P\left\{\begin{matrix}(1,2,3)\\3/(3, 1)\end{matrix}\right\} = \{(3, 1, 2)\}.$

7. Identified Permutation Theorem

Theorem 3: The number of permutations of N different components taken V at a time, whose first v components are identified, denoted by $P\left(\begin{matrix}N\\V/^vA\end{matrix}\right)$ is

$$P\left(\begin{matrix}N\\V/^vA\end{matrix}\right) = (N-v)(N-v-1)(N-v-2)\ ...\ (N-V+1)\ \underline{\hspace{1cm}}\quad(19)$$

Using the summation method we get (19) as

$$P\left(\begin{matrix}N\\V/^vA\end{matrix}\right) = \Sigma_{k_{v+1}=1}^{(N-v)}\Sigma_{k_{v+2}=1}^{(N-v-1)}\Sigma_{k_{v+3}=1}^{(N-v-2)}\ ...\ \Sigma_{k_V=1}^{(N-V+1)}C\ \underline{\hspace{0.8cm}}\quad(20)$$

Proof: We have given first v components are identified. Thus we fill the $(v+1)^{th}$ places with one of the $(N-v)$ components of the O. P. A. Now k_{v+1} states an index that indicates a component of the O. P. A. taken by $(v+1)^{th}$ place of a permutation holds the intervals

$$1 \le k_{v+1} \le k_1-1$$
$$k_1+1 \le k_{v+1} \le k_2-1$$
$$k_2+1 \le k_{v+1} \le k_3-1$$
$$\vdots \hspace{5cm}(21)$$
$$k_v+1 \le k_{v+1} \le N$$

where, $1 < k_1 < k_2 < k_3 < ... < N$. If the condition changes then the intervals will be changed. As $k_1, k_2, k_3, ..., k_v$ identified then suppose
$k_1 = 1, k_2 = 2, k_3 = 3, ..., k_v = v$.

Thus the particular interval becomes
$$v+1 \le k_{v+1} \le N.$$

The interval contains $(N-v)$ indexes. Thus the number of events characterizing $(v+1)^{th}$ components whose first v components are identified denoted by $S_{v+1/v}$ is the sum of numbers of indexes held by the interval (21) i.e.,

$$S_{v+1/v} = (N- v) \hspace{3cm} (22)$$

Using the summation method we get (22) as

$$S_{v+1/v} = \sum_{k_{v+1}=1}^{(N-v)} C \hspace{3cm} (23)$$

where C is a constant quantity taking unit value. It is memorize to you the change in $k_1, k_2, k_3, \dots, k_v$ does not effect in the number of indexes held in (21). After filling $(v+1)^{th}$ place with $k_{v+1}{}^{th}$ component, the $(v+2)^{th}$ place can be filled with one of the $(N-v-1)$ components of the O. P. A. Now k_{v+2} states an index that indicates a component of the O. P. A. taken by $(v+2)^{th}$ place of the permutation holds the intervals

$$1 \leq k_{v+2} \leq k_1-1$$
$$k_1+1 \leq k_{v+2} \leq k_2 -1$$
$$k_2 +1 \leq k_{v+2} \leq k_3 -1 \hspace{3cm} (24)$$
$$\vdots$$
$$k_v +1 \leq k_{v+2} \leq k_{v+1}-1$$
$$k_{v+1} +1 \leq k_{v+2} \leq N$$

where, $1< k_1 < k_2 < k_3 < \dots < N$. If the condition changes then the intervals will be changed. As $k_1, k_2, k_3, \dots, k_v$ are indexes that indicate the identified components thus they do not varies. Suppose

$$k_1= 1, \ k_2 = 2, \ k_3 = 3,\dots, k_v = v.$$

Then the particular intervals become when

$k_{v+1} = v+1$ then $\hspace{2cm}$ $v + 2 \leq k_{v+2} \leq N$; $N-v-1$

$k_{v+1} = v+2$ then $v+1 \leq k_{v+2} \leq v+1$ or, $v + 3 \leq k_{v+2} \leq N$; $1+N-v-2$

$k_{v+1} = v+3$ then $v+1 \leq k_{v+2} \leq v+2$ or, $v + 4 \leq k_{v+2} \leq N$; $2+N-v-3$

$\vdots$

$k_{v+1} = N$ then $v+1 \leq k_{v+2} \leq N-1$ or, $\hspace{2cm}$; $N-v-1$

The numbers after semicolons give the number of indexes held by the corresponding particular intervals. Thus the number of events of permutations characterizing $(v+2)^{th}$ components whose first v components are identified denoted by $S_{v+2/v}$ is the sum of numbers of indexes held by the intervals (24) i.e.,

$$S_{v+2/v} = (N-v-1) + (N-v-1) + (N-v-1) + \ldots\ldots + (N-v-1)$$

$$[(N-v) \text{ terms}]$$

$$= (N-v)(N-v-1) \qquad\qquad\qquad\qquad (25)$$

Using the summation method we get (25) as

$$S_{v+2/v} = \Sigma_{k_{v+1}=1}^{(N-v)} \Sigma_{k_{v+2}=1}^{(N-v-1)} C \qquad\qquad\qquad (26)$$

The change in $k_1, k_2, k_3, \ldots, k_v$ gives you change in the number of particular intervals for (24) but does not effect in the number of indexes held in (24). Again filling $(v+1)^{th}$ and $(v+2)^{th}$ places with $k_{v+1}{}^{th}$ and $k_{v+2}{}^{th}$ components respectively we can fill the $(v+3)^{th}$ place with one of the $(N-v-2)$ components of the O. P. A. Now k_{v+3} states an index that indicates a component of the O. P. A. taken by $(v+3)^{th}$ place of the permutation holds the intervals.

$$1 \le k_{v+3} \le k_1 - 1$$

$$k_1 + 1 \le k_{v+3} \le k_2 - 1$$

$$k_2 + 1 \le k_{v+3} \le k_3 - 1$$

$$\vdots \qquad\qquad\qquad\qquad\qquad\qquad (27)$$

$$k_v + 1 \le k_{v+3} \le k_{v+1} - 1$$

$$k_{v+1} + 1 \le k_{v+3} \le k_{v+2} - 1$$

$$k_{v+2} + 1 \le k_{v+3} \le N$$

where, $1 < k_1 < k_2 < k_3 < \ldots < N$. If the condition changes then the intervals will be changed. As $k_1, k_2, k_3, \ldots, k_v$ are indexes that indicates the identified components thus they do not varies. Suppose

$$k_1 = 1, \ k_2 = 2, \ k_3 = 3, \ldots, k_v = v.$$

then the particular intervals become when

$k_{v+1} = v+1$ and $k_{v+2} = v+2$ then $\qquad\qquad$ or, $v+3 \le k_{v+3} \le N$

$$; \ N-v-2$$

$k_{v+1} = v+1$ and $k_{v+2} = v+3$ then $v+2 \le k_{v+3} \le v+2$ or, $v+4 \le k_{v+3} \le N$

$$; \ 1 + N - v - 3$$

$k_{v+1} = v+1$ and $k_{v+2} = v+4$ then $v+2 \le k_{v+3} \le v+3$ or, $v+5 \le k_{v+3} \le N$

$$; \ 2 + N - v - 4$$

$$\vdots$$

$k_{v+1} = v+1$ and $k_{v+2} = N$ then $v+2 \le k_{v+3} \le N-1$ or, $\qquad\qquad$

$$; \ N-v-2$$

The numbers after semicolons give the number of indexes held by the corresponding particular intervals. Thus the number of events of permutations characterizing $(v+3)^{th}$ components whose first v components are identified, denoted by $S_{v+3/v}$, is the sum of numbers of indexes held by the intervals (16.4.42) i.e.,

$$S_{v+3/v} = (N-v-2) + (N-v-2) + (N-v-2) + \ldots\ldots + (N-v-2)$$

$$[(N-v)(N-v-1) \text{ terms}]$$

$$= (N-v)(N-v-1)(N-v-2) \underline{\hspace{3cm}} \qquad (28)$$

As $k_{v+1} = v+1, v+2, v+3, \ldots\ldots, N$.

Using the summation method we get (28) as

$$S_{v+3/v} = \Sigma_{k_{v+1}=1}^{(N-v)} \Sigma_{k_{v+2}=1}^{(N-v-1)} \Sigma_{k_{v+3}=1}^{(N-v-2)} C \underline{\hspace{2cm}} \qquad (29)$$

The change in $k_1, k_2, k_3, \ldots, k_v$ gives you change in the number of particular intervals for (27) but does not effect in the number of indexes held in (27). Continuing this process we get the number of events of permutations characterizing V^{th} components whose first v components are identified, denoted by $S_{V/v}$ is

$$S_{V/v} = (N-v)+(N-v-1)+(N-v-2)+\ldots\ldots+(N-V+1) \underline{\hspace{1cm}} \qquad (30)$$

Using the summation method we get (16.4.45) as

$$S_{V/v} = \Sigma_{k_{v+1}=1}^{(N-v)} \Sigma_{k_{v+2}=1}^{(N-v-1)} \Sigma_{k_{v+3}=1}^{(N-v-2)} \ldots\ldots \Sigma_{k_v=1}^{(N-V+1)} C \underline{\hspace{1cm}} \qquad (31)$$

As the last events are of one permutation, so the equations (30) and (31) give the number of permutations of N different components taken V at a time, whose first v components are identified.

Corollary 1: The number of permutations of N different components taken V at a time, whose first v components are identified, is the same as the number of permutations of (N−v) different components taken (V−v) at a time i.e.,

$$P\binom{N}{V/^vA} = P\binom{N-v}{V-v} \underline{\hspace{3cm}} \qquad (32)$$

Example 5: How many ways can an arrangement of 4 digits be chosen from 5 digits 1, 2, 3, 4 and 5 such that
(i) they are starting with 1
(ii) they are starting with 2

(iii) they are starting with 1 and 2

(iv) they are starting with 2 and 4.

Solution: Let the O. P. A. is ordered as

$1 \rightarrow 2 \rightarrow 3 \rightarrow 4 \rightarrow 5$

Now we have given $N = 5$ and $V = 4$.

(i) We fill the first place of the permutations with 1. Thus $k_1 = 1$.
Now from equation (19) we get

$$P\left(\begin{array}{c} 5 \\ 4/(1) \end{array}\right) = 4 \times 3 \times 2 = 24.$$

Again from equation (20) we get

$$P\left(\begin{array}{c} 5 \\ 4/(1) \end{array}\right) = \sum_{k_2=1}^{4} \sum_{k_3=1}^{3} \sum_{k_4=1}^{2} C$$

$$= \sum_{k_3=1}^{3} \sum_{k_4=1}^{2} C + \sum_{k_3=1}^{3} \sum_{k_4=1}^{2} C + \sum_{k_3=1}^{3} \sum_{k_4=1}^{2} C + \sum_{k_3=1}^{3} \sum_{k_4=1}^{2} C$$

$$= \left(\sum_{k_4=1}^{2} C + \sum_{k_4=1}^{2} C + \sum_{k_4=1}^{2} C\right) + \left(\sum_{k_4=1}^{2} C + \sum_{k_4=1}^{2} C + \sum_{k_4=1}^{2} C\right)$$

$$+ \left(\sum_{k_4=1}^{2} C + \sum_{k_4=1}^{2} C + \sum_{k_4=1}^{2} C\right) + \left(\sum_{k_4=1}^{2} C + \sum_{k_4=1}^{2} C + \sum_{k_4=1}^{2} C\right)$$

$$= (2C+2C+2C) + (2C+2C+2C) + (2C+2C+2C) + (2C+2C+2C)$$

$$= 6C + 6C + 6C + 6C$$

$$= 24C$$

$$= 24 \times 1 \qquad \text{[taking } C = 1]$$

$$= 24.$$

(ii) we fill the first place of the permutations with 2. Thus $k_1 = 2$.
Now from equation (19) we get

$$P\left(\begin{array}{c} 5 \\ 4/(2) \end{array}\right) = 4 \times 3 \times 2 = 24.$$

Again from equation (20) we get

$$P\left(\begin{array}{c} 5 \\ 4/(2) \end{array}\right) = \sum_{k_2=1}^{4} \sum_{k_3=1}^{3} \sum_{k_4=1}^{2} C = 24.$$

(iii) They are starting with 1 and 2. Thus $k_1 = 1$ and $k_2 = 2$.
Now the particular interval for k_3 is

$3 \leq k_3 \leq 5$

The interval contains 3 indexes. Thus

$S_{3/2} = 3$.

Again the particular intervals for k_4 are when

$k_3 = 3$ then ——— or, $4 \leq k_4 \leq 5$; 2

$k_3 = 4$ then $3 \leq k_4 \leq 3$ or, $5 \leq k_4 \leq 5$; 1+1

$k_3 = 5$ then $3 \le k_4 \le 4$ or,   ———   ; 2

Thus, $S_{4/2} = 3 \times 2 = 6$.

So the desired number is

$$P\binom{5}{4/(1,2)} = 6.$$

Using the summation method we get

$$\begin{aligned}
P\binom{5}{4/(1,2)} &= \sum_{k_3=1}^{3} \sum_{k_4=1}^{2} C \\
&= \sum_{k_4=1}^{2} C + \sum_{k_4=1}^{2} C + \sum_{k_4=1}^{2} C \\
&= 2C + 2C + 2C \\
&= 6C \\
&= 6 \times 1 \qquad \text{[taking } C = 1] \\
&= 6.
\end{aligned}$$

(iv) They are starting with 2 and 4. Thus $k_1 = 2$ and $k_2 = 4$.

Now the particular intervals for k_3 are

$1 \le k_3 \le 1$ or, $3 \le k_3 \le 3$ or, $5 \le k_3 \le 5$; $1 + 1 + 1$

Thus $S_{3/2} = 3$.

Again the particular intervals for k_3 are when

$k_3 = 1$ then ———   or, $3 \le k_4 \le 3$ or, $5 \le k_4 \le 5$; $1 + 1$

$k_3 = 3$ then $1 \le k_4 \le 1$ or, ———   or, $5 \le k_4 \le 5$; $1 + 1$

$k_3 = 5$ then $1 \le k_4 \le 1$ or, $3 \le k_4 \le 3$ or, ———   ; $1 + 1$

Thus $S_{4/2} = 6$.

So the desired number is

$$P\binom{5}{4/(2,4)} = 6$$

Using the summation method we get

$$P\binom{5}{4/(2,4)} = \sum_{k_3=1}^{3} \sum_{k_4=1}^{2} C = 6.$$

8. Permutation Partial Space

A permutation partial space is a set of all possible permutation (outcomes) of an experiment from a parent assembly (combination) C_t where the outcomes take order of the components into account. Let a permutation partial space contains T' possible outcomes then the permutation partial space denoted by $P\{C_t\}$ is

$P\{C_t\} = \{P_1, P_2, P_3, \cdots, P_t, \cdots, P_{T'}\}$ ———————————— (33)

Here every permutation contains V components.

Example 6: Find the permutation partial space of the experiment "arrange 3 digits 1, 2 and 3 taken all together.

Solution: We have given $A = (1, 2, 3) = C_t$ and $V = 3$.

Thus the permutation partial space is

$P\{(1, 2, 3)\} = \{(1, 2, 3), (1, 3, 2), (2, 1, 3), (2, 3, 1), (3, 1, 2), (3, 2, 1)\}.$

Theorem 4: The number of permutations of a parent combination $C_t = (C_{t1}, C_{t2}, C_{t3}, \ldots, C_{tv}, \ldots, C_{tV})$ denoted by $P(C_t)$ is

$P(C_t) = V!$ ———————————— (34)

Proof: We have given the parent combination $C_t = (C_{t1}, C_{t2}, C_{t3}, \ldots, C_{tv}, \ldots, C_{tV})$. It has V components that all different. Thus the components can be arranged in $V!$ ways taken all together. The theorem remembers you that of the number of permutations of V different objects taken all together.

Example 7: Find the number of permutations of a parent combination $C_t = (A, B, C, D, E)$.

Solution: We have given $C_t = (A, B, C, D, E)$ and $V = 5$.

So, $P((A, B, C, D, E)) = 5! = 120$.

9. Permutation Partial Event Type I

It is a set of all permutations (outcomes) of an experiment from an identified combination C_t where the outcomes take order of the components into account. Here the members are to be have same first components, same second components, same third components and so on same v^{th} components. A permutation partial event type I denoted by $P\{C_{tV/v}\}$ consists of $P_1, P_2, P_3, \cdots, P_t$, say, i.e.,

$P\{C_{tV/v}\} = \{P_1, P_2, P_3, \cdots, P_t\}$ ———————————— (35)

Example 8: Find the permutation partial events type I of the identified combinations

(i) $C_{t4/1} = (\underline{1}, 2, 3, 4)$ and (ii) $C_{t4/2} = (\underline{3,4}, 1, 2)$.

Solution: The partial events type I are

(i) $P\{(\underline{1}, 2, 3, 4)\} = \{(1, 2, 3, 4), (1, 2, 4, 3), (1, 3, 4, 2), (1, 3, 2, 4),$
$(1, 4, 2, 3), (1, 4, 3, 2)\}$

(ii) $P\{(\underline{3,4}, 1, 2)\} = \{(3, 4, 1, 2), (3, 4, 2, 1)\}$

Theorem 5: The number of permutations of an identified combination $C_{tV/v} = \left(\underline{C_{t1}, C_{t2}, C_{t3}, \ldots\ldots, C_{tv}}, \ldots\ldots, C_{tV}\right)$ whose first v components are identified, denoted by $P\left(C_{tV/v}\right)$ is

$$P\left(C_{tV/v}\right) = (V-v)! \hspace{3cm} (36)$$

Proof: We have given the identified combination $C_{tV/v} = \left(\underline{C_{t1}, C_{t2}, C_{t3}, \ldots\ldots, C_{tv}}, \ldots\ldots, C_{tV}\right)$ of V different components whose first v components are identified. Thus it has no possibility to change the places of the identified components. Hence the desired number of permutations is the number of permutations of rest $(V-v)$ components that not identified. Thus the number follows $(V-v)!$

Example 9: Find the number of permutations of the identified combination $C_{t5/2} = (\underline{A, B}, C, D, E)$ and find the permutation partial event type I.

Solution: We have given $C_{t5/2} = (\underline{A, B}, C, D, E)$, $V = 5$ and $v = 2$.

So, $P\left((\underline{A, B}, C, D, E)\right) = (5-2)! = 6$.

Now the permutation partial event type I is

$P\{(\underline{A, B}, C, D, E)\} = \{(A, B, C, D, E), (A, B, C, E, D), (A, B, D, C, E),$
$(A, B, D, E, C), (A, B, E, C, D), (A, B, E, D, C)\}$.

10. Permutation Partial Event Type II

It is a special kind of subset of a permutation partial space given an identified component assembly $^v A$ containing v identified components. Like definition (35) here the members are to be have same first components, same second components, same third components and so on same v^{th} components.

A permutation partial event type II denoted by $P\left\{ {C_t \atop V/^vA} \right\}$ consists of of $P_1, P_2, P_3, \cdots, P_t$, say , i.e.,

$$P\left\{ {C_t \atop V/^vA} \right\} = \{P_1, P_2, P_3, \cdots, P_t\} \quad\text{_____} \quad (37)$$

Example 10: Find the permutation partial events type II of the experiment " arrange 4 digits 1, 2, 3 and 4 taken all together" given that (i) $^1A = (3)$ and (ii) $^2A = (4, 1)$.

Solution: We have given $A = (1, 2, 3, 4) = C_t$, $V = 4$, $^1A = (3)$ and $^2A = (4, 1)$.

Hence the permutation partial events are

(i) $P\left\{ {(1, 2, 3, 4) \atop 4/(3)} \right\} = \{(3, 4, 1, 2), (3, 4, 2, 1), (3, 1, 2, 4), (3, 1, 4, 2),$ $(3, 2, 4, 1), (3, 2, 1, 4)\}$

(ii) $P\left\{ {(1, 2, 3, 4) \atop 4/(4, 1)} \right\} = \{(4, 1, 2, 3), (4, 1, 3, 2)\}$.

Theorem 6: The number of permutations occurring V components whose first v components are identified, of the parent combination C_t , denoted by $P\left({C_t \atop V/^vA} \right)$ is

$$P\left({C_t \atop V/^vA} \right) = (V-v) ! \quad\text{_____} \quad (38)$$

Proof: Let the parent combination $C_t = (C_{t1}, C_{t2}, C_{t3}, \ldots, C_{tv}, \ldots, C_{tV})$. We have the v identified components that belongs to anywhere in the parent combination we may arrange in a time the parent combination keeping v identified components in first. Thus the proof is follows as the theorem 5. Hence we get

$$P\left({C_t \atop V/^vA} \right) = (V-v) !$$

Example 11: Find the number of permutations of a combination $C_t =$ (1, 2, 3, 4, 5, 6, 7, 8) where identified components are 4, 7, 2 and 1.

Solution: We have given $C_t = (1, 2, 3, 4, 5, 6, 7, 8)$, $V = 8$, $v = 4$, $^4A = (4, 7, 2, 1)$.

So, $P\left(\begin{matrix}(1,2,3,4,5,6,7,8)\\8/\ (4,7,2,1)\end{matrix}\right) = (8-4)!\ = 24.$

Corollary 2: The number of permutations of an identified combination $C_{tV/v}$ denoted by $P\left(C_{tV/v}\right)$, the number of permutations occurring V components whose first v components are identified of the combination C_t denoted by $P\left(\begin{matrix}C_t\\V/^vA\end{matrix}\right)$ and the number of permutations of a parent combinations C'_t denoted by $P(C'_t)$ are the same i.e.,

$$P\left(C_{tV/v}\right) = P\left(\begin{matrix}C_t\\V/^vA\end{matrix}\right) = P(C'_t) \qquad\qquad (39)$$

$$\text{where},\ C'_t = \left(C'_{t1}, C'_{t2}, C'_{t3}, \ldots\ldots, C'_{tv'}\right)$$

$$V' = (V-v)$$

11. General Permutation Theorem

Theorem 7: The number of permutations in a permutation space of N different components taken V at a time where there occurred Y particular components taken from $U \le V$ particular components (limited size of components) that occurred in the parent component assembly of N components denoted by $P\left(\begin{matrix}N & U\\V & Y\end{matrix}\right)$ is

$$P\left(\begin{matrix}N & U\\V & Y\end{matrix}\right) = V!\ C\left(\begin{matrix}U\\Y\end{matrix}\right) C\left(\begin{matrix}N-U\\V-Y\end{matrix}\right) \qquad\qquad (40)$$

$$\text{where},\ N \ge V$$

$$U \le V$$

$$k \le Y \le U$$

$$0 \le k = U - N + V$$

Proof: Let the permutation space is of N components taken V at a time. We can select Y components out of U components without restriction is $C\left(\begin{matrix}U\\Y\end{matrix}\right)$. Now for each combination we can select $C\left(\begin{matrix}N-U\\V-Y\end{matrix}\right)$ combinations. Again for the total combinations each combination can be arranged in V! ways. Thus finally we get the total number of permutations is

$$V!\ C\left(\begin{matrix}U\\Y\end{matrix}\right) C\left(\begin{matrix}N-U\\V-Y\end{matrix}\right).$$

Example 12: Let the parent component assembly M = (A, B, C, D, E). Find the number of permutations of the following and then write the permutations.

(i) $P\begin{pmatrix} 5 & 3 \\ 3 & 3 \end{pmatrix}$, (ii) $P\begin{pmatrix} 5 & 3 \\ 3 & 2 \end{pmatrix}$, (iii) $P\begin{pmatrix} 5 & 3 \\ 3 & 1 \end{pmatrix}$, (iv) $P\begin{pmatrix} 5 & 2 \\ 3 & 2 \end{pmatrix}$,

(v) $P\begin{pmatrix} 5 & 2 \\ 3 & 1 \end{pmatrix}$, (vi) $P\begin{pmatrix} 5 & 2 \\ 3 & 0 \end{pmatrix}$, (vii) $P\begin{pmatrix} 5 & 1 \\ 3 & 1 \end{pmatrix}$, (viii) $P\begin{pmatrix} 5 & 1 \\ 3 & 0 \end{pmatrix}$

Solution:

(i) $P\begin{pmatrix} 5 & 3 \\ 3 & 3 \end{pmatrix} = 3! \, C\begin{pmatrix} 3 \\ 3 \end{pmatrix} C\begin{pmatrix} 5-3 \\ 3-3 \end{pmatrix} = 6 \times 1 \times 1 = 6.$

Now the permutations are (A, B, C), (A, C, B), (B, A, C), (B, C, A), (C, A, B), (C, B, A).

(ii) $P\begin{pmatrix} 5 & 3 \\ 3 & 2 \end{pmatrix} = 3! \, C\begin{pmatrix} 3 \\ 2 \end{pmatrix} C\begin{pmatrix} 5-3 \\ 3-2 \end{pmatrix} = 6 \times 3 \times 2 = 36.$

Now the permutations are (A, B, D), (A, D, B), (B, A, D), (B, D, A), (D, A, B), (D, B, A), (A, B, E), (A, E, B), (B, A, E), (B, E, A), (E, A, B), (E, B, A), (A, C, D), (A, D, C), (C, A, D), (C, D, A), (D, A, C), (D, C, A), (A, C, E), (A, E, C), (C, A, E), (C, E, A), (E, A, C), (E, C, A), (B, C, D), (B, D, C), (C, B, D), (C, D, B), (D, B, C), (D, C, B), (B, C, E), (B, E, C), (C, B, E), (C, E, B), (E, B, C), (E, C, B).

(iii) $P\begin{pmatrix} 5 & 3 \\ 3 & 1 \end{pmatrix} = 3! \, C\begin{pmatrix} 3 \\ 1 \end{pmatrix} C\begin{pmatrix} 5-3 \\ 3-1 \end{pmatrix} = 6 \times 3 \times 1 = 18.$

Now the permutations are (A, D, E), (A, E, D), (D, A, E), (D, E, A), (E, A, D), (E, D, A), (B, D, E), (B, E, D), (D, B, E), (D, E, B), (E, B, D), (E, D, B), (C, D, E), (C, E, D), (D, C, E), (D, E, C), (E, C, D), (E, D, C).

(iv) $P\begin{pmatrix} 5 & 2 \\ 3 & 2 \end{pmatrix} = 3! \, C\begin{pmatrix} 2 \\ 2 \end{pmatrix} C\begin{pmatrix} 5-2 \\ 3-2 \end{pmatrix} = 6 \times 1 \times 3 = 18.$

Now the permutations are (A, B, C), (A, C, B), (B, A, C), (B, C, A), (C, A, B), (C, B, A), (A, B, D), (A, D, B), (B, A, D), (B, D, A), (D, A, B), (D, B, A), (A, B, E), (A, E, B), (B, A, E), (B, E, A), (E, A, B), (E, B, A).

(v) $P\begin{pmatrix} 5 & 2 \\ 3 & 1 \end{pmatrix} = 3! \, C\begin{pmatrix} 2 \\ 1 \end{pmatrix} C\begin{pmatrix} 5-2 \\ 3-1 \end{pmatrix} = 6 \times 2 \times 3 = 36.$

Now the permutations are (A, C, D), (A, D, C), (C, A, D), (C, D, A), (D, A, C), (D, C, A), (A, C, E), (A, E, C), (C, A, E), (C, E, A), (E, A, C), (E, C, A), (A, D, E), (A, E, D), (D, A, E), (D, E, A), (E, A, D), (E, D, A), (B, C, D), (B, D, C), (C, B, D), (C, D, B), (D, B, C), (D, C, B), (B, C, E), (B, E, C), (C, B, E), (C, E, B), (E, B, C), (E, C, B), (B, D, E), (B, E, D), (D, B, E), (D, E, B), (E, B, D), (E, D, B).

(vi) $P\binom{5\ \ 2}{3\ \ 0} = 3\ !\ C\binom{2}{0}C\binom{5-2}{3-0} = 6 \times 1 \times 1 = 6.$

Now the permutations are (C, D, E), (C, E, D), (D, C, E), (D, E, C), (E, C, D), (E, D, C).

(vii) $P\binom{5\ \ 1}{3\ \ 1} = 3\ !\ C\binom{1}{1}C\binom{5-1}{3-1} = 6 \times 1 \times 6 = 36.$

Now the permutations are (A, B, C), (A, C, B), (B, A, C), (B, C, A), (C, A, B), (C, B, A), (A, B, D), (A, D, B), (B, A, D), (B, D, A), (D, A, B), (D, B, A), (A, B, E), (A, E, B), (B, A, E), (B, E, A), (E, A, B), (E, B, A), (A, C, D), (A, D, C), (C, A, D), (C, D, A), (D, A, C), (D, C, A), (A, C, E), (A, E, C), (C, A, E), (C, E, A), (E, A, C), (E, C, A), (A, D, E), (A, E, D), (D, A, E), (D, E, A), (E, A, D), (E, D, A).

(viii) $P\binom{5\ \ 1}{3\ \ 0} = 3\ !\ C\binom{1}{0}C\binom{5-1}{3-0} = 6 \times 1 \times 4 = 24.$

Now the permutations are (B, C, D), (B, D, C), (C, B, D), (C, D, B), (D, B, C), (D, C, B), (B, C, E), (B, E, C), (C, B, E), (C, E, B), (E, B, C), (E, C, B), (B, D, E), (B, E, D), (D, B, E), (D, E, B), (E, B, D), (E, D, B), (C, D, E), (C, E, D), (D, C, E), (D, E, C), (E, C, D), (E, D, C).

Now we discuss the probability law in which variables of frequency distribution got from a permutation space. In this case we get a theoretical discrete distribution specially called permutation distribution.

12. Permutation Distribution

A random variable Y is said to follow permutation distribution if it assumes only non-negative values and its probability mass function is given by

$$P(Y) = P(Y;\ N,\ U,\ V) = \frac{V\ !\ C\binom{U}{Y}C\binom{N-U}{V-Y}}{P\binom{N}{V}}\ ;\ Y = k\ ,\ k+1,\ k+2\ ,........,\ U$$

$$0 \le k = U-N+V$$

$$= 0;\ \text{otherwise.} \ \underline{\hspace{3cm}} \tag{41}$$

The three independent finite constants N, U and V are known as the parameters of this distribution. Permutation distribution is a discrete distribution as Y can take only the non-negative values under the interval $U-N+V \le Y \le U$. Any variable which follows permutation distribution is known as permutation variate and denoted by the symbol $Y \sim P(N, U, V)$.

Remark: It should be noted that

$$\Sigma_Y P(Y \; ; \; N, U, V) = \Sigma_Y \frac{V! \, C\binom{U}{Y} \, C\binom{N-U}{V-Y}}{P\binom{N}{V}} = 1$$

$$\Rightarrow \Sigma_Y V! \, C\binom{U}{Y} \, C\binom{N-U}{V-Y} = P\binom{N}{V} \qquad\qquad\qquad (42)$$

Example 13: Let the parent component assembly M = (A, B, C, D, E). Find the probability of permutations taken 4 at a time in which they have

(i) first 3 components

(ii) exactly any 2 of first 3 components

(iii) the components C and D

(iv) exactly any 1 of the components C and D.

Solution: (i) We have given the parameters N = 5, U = 3 and V = 4. The probability of Y = 3 is given by

$$P(3 \; ; \; 5, 3, 4) = \frac{4! \, C\binom{3}{3} C\binom{5-3}{4-3}}{P\binom{5}{4}} = \frac{48}{120} = 0.4$$

(ii) We have given the parameters N = 5, U = 3 and V = 4. The probability of Y = 2 is given by

$$P(2 \; ; \; 5, 3, 4) = \frac{4! \, C\binom{3}{2} C\binom{5-3}{4-2}}{P\binom{5}{4}} = \frac{72}{120} = 0.6$$

(iii) We have given the parameters N = 5, U = 2 and V = 4. The probability of Y = 2 is given by

$$P(2 \; ; \; 5, 2, 4) = \frac{4! \, C\binom{2}{2} C\binom{5-2}{4-2}}{P\binom{5}{4}} = \frac{72}{120} = 0.6$$

(iv) We have given the parameters N = 5, U = 2 and V = 4. The probability of getting Y = 1 is given by

$$P(1 \; ; \; 5, 2, 4) = \frac{4! \, C\binom{2}{1} C\binom{5-2}{4-1}}{P\binom{5}{4}} = \frac{48}{120} = 0.4.$$

10.1 Moments:

The first four moments about origin of permutation distribution are obtained as follows:

$$\mu_1' = E(Y) = \Sigma_Y Y \, P(Y \; ; \; N, U, V) = \Sigma_Y Y V! \, C\binom{U}{Y} \, C\binom{N-U}{V-Y} \Big/ P\binom{N}{V}$$

$$= \Sigma_Y Y V! \, \frac{U!}{Y!(U-Y)!} C\binom{N-U}{V-Y} \Big/ P\binom{N}{V}$$

$$= \frac{UV!}{P\binom{N}{V}} \sum_Y C\binom{U-1}{Y-1} C\binom{N-U}{V-Y}$$

$$= \frac{UV!}{C\binom{N}{V}} C\binom{N-1}{V-1} = \frac{UV}{N}.$$

$$\mu_2' = E(Y^2) = \sum_Y Y^2 \, P(Y\,;\,N,U,V)$$

$$= \sum_Y Y^2 \, V! \, C\binom{U}{Y} C\binom{N-U}{V-Y} / P\binom{N}{V}$$

$$= \sum_Y \{Y(Y-1) + Y\} \, V! \, C\binom{U}{Y} C\binom{N-U}{V-Y} / P\binom{N}{V}$$

$$= \frac{U(U-1)V!}{P\binom{N}{V}} \sum_Y \left\{1 + \frac{1}{Y-1}\right\} C\binom{U-2}{Y-2} C\binom{N-U}{V-Y}$$

$$= \frac{U(U-1)V!}{P\binom{N}{V}} \left[\sum_Y C\binom{U-2}{Y-2} C\binom{N-U}{V-Y} + \sum_Y \frac{1}{Y-1} C\binom{U-2}{Y-2} C\binom{N-U}{V-Y} \right]$$

$$= \frac{U(U-1)V!}{P\binom{N}{V}} \times C\binom{N-2}{V-2} + \frac{UV!}{P\binom{N}{V}} \times C\binom{N-1}{V-1}$$

$$= \frac{UV(U-1)(V-1)}{N(N-1)} + \frac{UV}{N}.$$

$$\mu_3' = E(Y^3) = \sum_Y Y^3 \, P(Y\,;\,N,U,V) = \sum_Y Y^3 \, V! \, C\binom{U}{Y} C\binom{N-U}{V-Y} / P\binom{N}{V}$$

$$= \sum_Y \{Y(Y-1)(Y-2) + 3Y(Y-1) + Y\} C\binom{U}{Y} C\binom{N-U}{V-Y} / P\binom{N}{V}$$

$$= \frac{U(U-1)(U-2)V!}{P\binom{N}{V}} \sum_Y \left\{1 + \frac{3}{(Y-2)} + \frac{1}{(Y-1)(Y-2)}\right\} C\binom{U-3}{Y-3} C\binom{N-U}{V-Y}$$

$$= \frac{U(U-1)(U-2)V!}{P\binom{N}{V}} C\binom{N-3}{V-3} + \frac{3U(U-1)V!}{P\binom{N}{V}} C\binom{N-2}{V-2} + \frac{UV!}{P\binom{N}{V}} C\binom{N-1}{V-1}$$

$$= \frac{UV(U-1)(U-2)(V-1)(V-2)}{N(N-1)(N-2)} + \frac{3UV(U-1)(V-1)}{N(N-1)} + \frac{UV}{N}.$$

$$\mu_4' = E(Y^4) = \sum_Y Y^4 \, P(Y\,;\,N,U,V) = \sum_Y Y^4 \, V! \, C\binom{U}{Y} C\binom{N-U}{V-Y} / P\binom{N}{V}$$

$$= \sum_Y \{Y(Y-1)(Y-2)(Y-3) + 6Y(Y-1)(Y-2) + 7Y(Y-1) + Y\}$$
$$\times C\binom{U}{Y} C\binom{N-U}{V-Y} / P\binom{N}{V}$$

$$= \frac{U(U-1)(U-2)(U-3)V!}{P\binom{N}{V}} \sum_Y \left\{1 + \frac{6}{(Y-3)} + \frac{7}{(Y-2)(Y-3)} + \frac{1}{(Y-1)(Y-2)(Y-3)}\right\}$$
$$\times C\binom{U-4}{Y-4} C\binom{N-U}{V-Y}$$

$$= \frac{U(U-1)(U-2)(U-3)V!}{P\binom{N}{V}} C\binom{N-4}{V-4} + \frac{6U(U-1)(U-2)V!}{P\binom{N}{V}} C\binom{N-3}{V-3}$$

$$+ \frac{7U(U-1)V!}{P\binom{N}{V}} C\binom{N-2}{V-2} + \frac{UV!}{P\binom{N}{V}} C\binom{N-1}{V-1}$$

$$= \frac{U(U-1)(U-2)(U-3)V(V-1)(V-2)(V-3)}{N(N-1)(N-2)(N-3)} + \frac{6U(U-1)(U-2)V(V-1)(V-2)}{N(N-1)(N-2)}$$
$$+ \frac{7U(U-1)V(V-1)}{N(N-1)} + \frac{UV}{N}.$$

Now variance $\mu_2 = \mu_2' - \mu_1'^2$

$$= \frac{UV(U-1)(V-1)}{N(N-1)} + \frac{UV}{N} - \frac{U^2V^2}{N^2}$$

$$= \frac{NUV(U-1)(V-1) + NUV(N-1) - U^2V^2(N-1)}{N^2(N-1)}$$

$$= \frac{NU^2V^2 - NU^2V - NUV^2 + NUV + N^2UV - NUV - NU^2V^2 + U^2V^2}{N^2(N-1)}$$

$$= \frac{-NU^2V - NUV^2 + N^2UV - NU^2V^2 + U^2V^2}{N^2(N-1)}$$

$$= \frac{UV(N-U)(N-V)}{N^2(N-1)}.$$

10.2: Factorial moments of this distribution:

The kth factorial moment of this distribution is obtained as follows:

$$\mu'_{(k)} = E\{Y^{(k)}\} = \sum_Y Y^{(k)} \frac{V! C\binom{U}{Y} C\binom{N-U}{V-Y}}{P\binom{N}{V}}$$

$$= U^{(k)} \sum_Y \frac{V! C\binom{U-k}{Y-k} C\binom{N-U}{V-Y}}{P\binom{N}{V}} = U^{(k)} \frac{C\binom{N-k}{V-k}}{P\binom{N}{V}} = \frac{U^{(k)}V^{(k)}}{N^{(k)}}.$$

Now, $\mu'_{(1)} = E\{Y^{(1)}\} = \frac{U^{(1)}V^{(1)}}{N^{(1)}} = \frac{UV}{N}$

$\mu'_{(2)} = E\{Y^{(2)}\} = \frac{U^{(2)}V^{(2)}}{N^{(2)}} = \frac{U(U-1)V(V-1)}{N(N-1)} = \frac{UV(U-1)(V-1)}{N(N-1)}$

$\mu'_{(3)} = E\{Y^{(3)}\} = \frac{U^{(3)}V^{(3)}}{N^{(3)}} = \frac{U(U-1)(U-2)V(V-1)(V-2)}{N(N-1)(N-2)}$

$$= \frac{UV(U-1)(U-2)(V-1)(V-2)}{N(N-1)(N-2)}.$$

10.3: Mode of this distribution:

We have, $\frac{P(Y;N,U,V)}{P(Y-1;N,U,V)} = \frac{V! C\binom{U}{Y} C\binom{N-U}{V-Y}/P\binom{N}{V}}{V! C\binom{U}{Y-1} C\binom{N-U}{V-Y+1}/P\binom{N}{V}}$

i.e., $= \frac{P(Y)}{P(Y-1)} = \frac{UV - UY - VY + V^2 - 2Y + U + V + 1}{NY - UY - VY + Y^2}$

Now if $\frac{P(Y)}{P(Y-1)} > 1$

then, $UV - UY - VY + V^2 - 2Y + U + V + 1 > NY - UY - VY + Y^2$

or, $Y < \frac{(U+1)(V+1)}{(N+2)}$

and if $\dfrac{P(Y)}{P(Y-1)} < 1$

then , $Y > \dfrac{(U+1)(V+1)}{(N+2)}$

We discuss the following two cases:

Case I: when $\dfrac{(U+1)(V+1)}{(N+2)}$ is not an integer.

Let $\dfrac{(U+1)(V+1)}{(N+2)} = t + f$; where t is an integer and f is a fractional such that $0 < f < 1$.

then we get $\dfrac{P(Y)}{P(Y-1)} > 1$; for $Y = k , k+1,\ k+2,\ \ldots\ldots, t$

$$0 \le k = U - N + V$$

and $\dfrac{P(Y)}{P(Y-1)} < 1$; for $Y = t+1, t+2, t+3, \ldots\ldots, U$

$\Rightarrow \dfrac{P(k+1)}{P(k)} > 1 , \quad \dfrac{P(k+2)}{P(k+1)} > 1 , \ldots\ldots, \dfrac{P(t)}{P(t-1)} > 1$

and $\dfrac{P(t+1)}{P(t)} < 1, \quad \dfrac{P(t+2)}{P(t+1)} < 1 , \ldots\ldots, \dfrac{P(U)}{P(U-1)} < 1.$

Thus $P(k) < P(k+1) < P(k+2) \ldots\ldots < P(t-1) < P(t) > P(t+1) > P(t+2) > \ldots\ldots > P(U)$.

In this case we have one maximum value and this is P(t). Thus the modal value is t, the

integral part of $\dfrac{(U+1)(V+1)}{(N+2)}$.

Case II: When $\dfrac{(U+1)(V+1)}{(N+2)}$ is an integer.

Let $\dfrac{(U+1)(V+1)}{(N+2)} = t$, an integer then we get

$\dfrac{P(Y)}{P(Y-1)} > 1$; for $Y = k , k+1,\ k+2,\ \ldots\ldots, t-1$

$$0 \le k = U - N + V$$

$\dfrac{P(Y)}{P(Y-1)} = 1$; for $Y = t$

and $\dfrac{P(Y)}{P(Y-1)} < 1$; for $Y = t+1 , t+2 , \ldots\ldots, U$

Now proceeding as case I we get

$P(k) < P(k+1) < P(k+2) \ldots\ldots < P(t-1) = P(t) > P(t+1) > P(t+2) > \ldots\ldots > P(U)$

In this case we get two maximum values i.e., P(t−1) and P(t). Thus the modal values are (t−1) and t i,e., $\left\{ \dfrac{(U+1)(V+1)}{(N+2)} - 1 \right\}$ and $\dfrac{(U+1)(V+1)}{(N+2)}$.

10.4: Recurrence relation for this distribution:

We have $P(Y) = V! \, C\binom{U}{Y} \, C\binom{N-U}{V-Y} / P\binom{N}{V}$

and $P(Y+1) = V! \, C\binom{U}{Y+1} \, C\binom{N-U}{V-Y-1} / P\binom{N}{V}$

Now, $\dfrac{P(Y+1)}{P(Y)} = \dfrac{C\binom{U}{Y+1} C\binom{N-U}{V-Y-1}/P\binom{N}{V}}{C\binom{U}{Y} C\binom{N-U}{V-Y}/P\binom{N}{V}} = \dfrac{(U-Y)(V-Y)}{(Y+1)(N-U-V+Y+1)}$

So, $P(Y+1) = \left\{ \dfrac{(U-Y)(V-Y)}{(Y+1)(N-U-V+Y+1)} \right\} P(Y)$.

This is the required recurrence relation.

10.5: Combination distribution from this distribution:

We have $P(Y\,;\,N, U, V) = V! \, C\binom{U}{Y} \, C\binom{N-U}{V-Y} / P\binom{N}{V}$

$$= \frac{V! \, C\binom{U}{Y} \, C\binom{N-U}{V-Y}}{\frac{N!}{(N-V)!}}$$

Dividing both numerator and denominator of the right side of the above equation we get

$$P(Y\,;\,N, U, V) = \frac{\frac{V! \, C\binom{U}{Y} \, C\binom{N-U}{V-Y}}{V!}}{\frac{N!}{(N-V)! \, V!}} = \frac{C\binom{U}{Y} \, C\binom{N-U}{V-Y}}{C\binom{N}{V}} = C(Y\,;\,N, U, V)$$

which is the probability function of the combination distribution with parameter N, U and V.

10.6 : Binomial Distribution from this distribution:

To perform this work permutation distribution changes into combination distribution followed by 10.6 and then changes into binomial distribution.

Example 14: Find the probability of permutations from 8 digits 1, 2, 3, 4, 5, 6, 7 and 8 taken 4 at a time in which they have

(i) the digits 2, 4 and 5

(ii) any two of the digits 2, 4 and 5

(iii) the components 6 and 7

(iv) any one of the digits 6 and 7

(v) the digits 1, 3, 5 and 7.

Solution: (i) We have given the parameters N = 8, U = 3, V = 4 and Y = 3.

Thus the number of favourable cases is

$$P\binom{8 \; 3}{4 \; 3} = 4! \, C\binom{3}{3} C\binom{8-3}{4-3} = 120.$$

But the number of possible cases is

$$P\binom{8}{4} = 1680.$$

Thus the probability of getting $Y = 3$ is

$$P(3 \; ; 8, 3, 4) = \frac{120}{1680} = 0.07$$

(ii) We have given the parameters $N = 8$, $U = 3$, $V = 4$ and $Y = 2$.

Thus the number of favourable cases is

$$P\binom{8 \; 3}{4 \; 2} = 4! \, C\binom{3}{2} C\binom{8-3}{4-2} = 720.$$

But the number of possible cases is

$$P\binom{8}{4} = 1680$$

Thus the probability of getting $Y = 2$ is

$$P(2 \; ; 8, 3, 4) = \frac{720}{1680} = 0.42$$

(iii) We have given the parameters $N = 8$, $U = 2$, $V = 4$ and $Y = 2$.

Thus the number of favourable cases is

$$P\binom{8 \; 2}{4 \; 2} = 4! \, C\binom{2}{2} C\binom{8-2}{4-2} = 360.$$

But the number of possible cases is

$$P\binom{8}{4} = 1680$$

Thus the probability of getting $Y = 2$ is

$$P(2 \; ; 8, 2, 4) = \frac{360}{1680} = 0.214$$

(iv) We have given the parameters $N = 8$, $U = 2$, $V = 4$ and $Y = 1$.

Thus the number of favourable cases is

$$P\binom{8 \; 2}{4 \; 1} = 4! \, C\binom{2}{1} C\binom{8-2}{4-1} = 960.$$

But the number of possible cases is

$$P\binom{8}{4} = 1680$$

Thus the probability of getting $Y = 3$ is

$$P(1 \; ; 8, 2, 4) = \frac{960}{1680} = 0.571$$

(v) We have given the parameters $N = 8$, $U = 4$, $V = 4$ and $Y = 4$.

Thus the number of favourable cases is

$$P\binom{8 \; 4}{4 \; 4} = 4! \, C\binom{4}{4} C\binom{8-4}{4-4} = 24.$$

But the number of possible cases is

$$P\binom{8}{4} = 1680$$

Thus the probability of getting $Y = 4$ is

$$P(4 ; 8, 4, 4) = \frac{24}{1680} = 0.0142.$$

13. Selected Permutations

The permutation theorem for an assembly of N components that has some alike components is large different from the theorem 16.4.1. Here the first theorem introduces the number of permutations when the components are not all different. Suppose it has N_1 components alike of one kind, N_2 components alike f another kind, N_3 components alike of a third kind and so on for a hth kind then there called first kind of components occupies from first place, second kind of components occupies from $(N_1+1)^{th}$ place, third kind of components occupies from $(N_1+N_2+1)^{th}$ place and so on h^{th} kind of components occupies from $(N_1+N_2+N_3+ ... +N_{h-1}+1)^{th}$ place in the row where $N_1+N_2+N_3+ ... +N_h = N$. In this case we omit the other places.

14. Selected Permutation Theorem

Theorem 8: The number of permutations of N components that not all different and there are N_1 components alike of one kind, N_2 components alike of another kind, N_3 components alike of a third kind and so on for a hth kind, taken V at a time, denoted by $P^* \binom{N}{V}$ is

$$P^* \binom{N}{V} = \sum_{g=1}^{S_{V-1}^*} S_{V.g}^* \quad \rule{3cm}{0.4pt} \quad (43)$$

Using the summation method we get (43) as

$$P^* \binom{N}{V} = \sum_{k_1^* \in K_1^*} \sum_{k_2^* \in K_2^*} \sum_{k_3^* \in K_3^*} \cdots \cdots \sum_{k_V^* \in K_V^*} C \quad \rule{2cm}{0.4pt} \quad (44)$$

where C is a constant quantity taking unit value.

Proof: Suppose we have N components that not all different. Let there are N_1 components alike of one kind, N_2 components alike of another kind, N_3 components alike of a third kind and so on N_h components alike of a hth kind then we have k_1 the index that indicates a components of the O. P. A. as,

$$1 \leq k_1 \leq N \qquad \text{[See (6)]}$$

and k_1^* the index that indicates the component of the O. P. A. which is not alike of previous components indicated by previous indexes of the interval (6) as

$1 \leq k_1^* \leq$ till N ──────────────── (45)

Clearly the selected indexes are i.e., $k*_1$ takes

$1, (N_1+ 1), (N_1+N_2+1)$ .... till N

Thus the assembly of assemblies of selected indexes is

$K_1^* = (K_{11}^*)$ ──────────────── (46)

where , $K_{11}^* = (1, (N_1+1), (N_1+N_2+1)$ .... till N)

The number of selected events characterizing first components denoted by S_1^* is the number of indexes held by the assembly of assemblies (46) i.e.,

$S_1^* = S_{11}^* = \sum_{g=1}^{S_0^*} S_{1.g}^*$ ──────────────── (47)

where , $S_0^* = 1$

and $S_{11}^* =$ number of indexes held by the assembly K_{11}^*

Using the summation method we get (47) as

$S_1^* = \sum_{k_1^* \in K_1^*} C$ ──────────────── (48)

Again we have k_2, the index that indicates a component of the O. P. A. taken by second place of a permutation as

$1 \leq k_2 \leq k_1 - 1$

$k_1 + 1 \leq k_2 \leq N$ [see (9)]

where $1 < k_1 < N$. If the condition changes then the intervals will be changed.

Now k_2^*, the index that indicates the component of the O. P. A. which is not alike of previous components indicated by previous indexes contained in the interval (9) as

$1 \leq k_2^* \leq$ till $(k_1^* -1)$

$k_1^* + 1 \leq k_2^* \leq$ till N ──────────────── (49)

where $1 < k_1^* <$ till N

Clearly the selected indexes are i.e., when

$k_1^* = 1$ then $k_2^* =$ ──────── or, 2 , ... till N

$k_1^* = (N_1+ 1)$ then $k_2^* = 1$, ..., till N_1 or, $N_1+ 2$ .... till N

$k_1^* = (N_1+N_2+1)$ then $k_2^* = 1$,........., till (N_1+N_2)

or, (N_1+N_2+2), ..., till N

and so on.

The dots takes only selected indexes.

Thus the assembly of assemblies of selected indexes is

$$K_2^* = (K_{21}^*, K_{22}^*, K_{23}^*, \ldots, K_{2.S_1^*}^*) \quad\text{————————}\quad (50)$$

$$\text{where}, \ K_{21}^* = (2,\ldots, \text{till } N)$$

$$K_{22}^* = (1,\ldots, \text{till } N_1, (N_1+2), \ldots, \text{till } N)$$

$$K_{23}^* = (1,\ldots, \text{till } (N_1+N_2), (N_1+N_2+2)\ldots, \text{till } N)$$

and so on.

So the number of selected events characterizing second components denoted by S_2^* is the sum of numbers of indexes held by the assembly of assemblies (50) i.e.,

$$S_2^* = S_{21}^* + S_{22}^* + S_{23}^* + \ldots S_1^* \text{ terms.}$$

$$= \sum_{g=1}^{S_1^*} S_{2.g}^* \quad\text{————————}\quad (51)$$

$$\text{where}, \ S_{21}^* = \text{number of indexes held by the assembly } K_{21}^*$$

$$S_{22}^* = \text{number of indexes held by the assembly } K_{22}^*$$

$$S_{23}^* = \text{number of indexes held by the assembly } K_{23}^*$$

and so on.

Using the summation method we get (51) as

$$S_2^* = \sum_{k_1^* \in K_1^*} \sum_{k_2^* \in K_2^*} C \quad\text{————————}\quad (52)$$

Again we have k_3, the index that indicates a component of the O. P. A. taken by third place of the permutation as

$$1 \leq k_3 \leq k_1-1$$

$$k_1+1 \leq k_3 \leq k_2 -1 \qquad [\text{see } (12)]$$

$$k_2 +1 \leq k_3 \leq N$$

where, $1 < k_1 < k_2 < N$, If the condition changes then the intervals will be changed.

Now k_3^*, the index that indicates the component of the O. P. A. which is not alike of previous component indicated by previous indexes contained in the intervals (12) as

$$1 \leq k_3^* \leq \text{till } k_1^*-1$$

$$k_1^*+1 \leq k_3^* \leq \text{till } k_2^*-1$$

$$k_2^*+1 \leq k_3^* \leq \text{till } N$$

where, $1 < k_1^* < k_2^* < N$

Clearly the selected indexes are i.e., when

$k_1^* = 1$ and $k_2^* = 2$ then k_3^* takes $3,\ldots$ till N

$\vdots$

$k_1^* = 1$ and $k_2^* = $ till N then k_3^* takes

$$2,........, \text{till } (N_1+N_2+N_3+ .... + N_{h-1})$$

$k_1^* = (N_1+1)$ and $k_2^* = 1$ then k_3^* takes _____ or, _____

or, $(N_1+2), ..., $ till N

$\vdots$

$k_1^* = (N_1+1)$ and $k_2^* = $ till N_1 then k_3^* takes _____ or, _____

or, $(N_1+2), ..., $ till N

$k_1^* = (N_1+1)$ and $k_2^* = (N_1+2)$ then k_3^* takes $1, ..., $ till N_1

or, _____ or, $(N_1+3), ..., $ till N

$\vdots$

$k_1^* = (N_1+1)$ and $k_2^* = $ till N then k_3^* takes $1 ,...., $ till N_1

or, $(N_1+2), ..., $ till $(N_1+N_2+N_3+ .... + N_{h-1})$ or, _____

$k_1^* = (N_1+N_2+1)$ and $k_2^* = 1$ then k_3^* takes _____

or, $2, ..., $ till (N_1+N_2) or, $(N_1+N_2+2), ..., $ till N

$\vdots$

$k_1^* = (N_1+N_2+1)$ and $k_2^* = $ till (N_1+N_2) then k_3^* takes $1 ,........, $ till N_1

or, _____ or, $(N_1+N_2+2) ,......, $ till N

$k_1^* = (N_1+N_2+1)$ and $k_2^* = (N_1+N_2+2)$ then k_3^* takes

$1 ,........, $ till (N_1+N_2) or, _____ or, $(N_1+N_2+3) ,......, $ till N

$\vdots$

$k_1^* = (N_1+N_2+1)$ and $k_2^* = $ till N then k_3^* takes $1,........, $ till (N_1+N_2)

or, $(N_1+N_2+2) ,......, $ till $(N_1+N_2+N_3+ .... +N_{h-1})$ or, _____

and so on.

The dots after comma takes only selected indexes. Similarly we can find the selected indexes for $K_1^* = (N_1+N_2+N_3+1), (N_1+N_2+N_3+N_4+1)$ etc.

Here the assembly of assemblies of selected indexes is

$$K_3^* = (K_{31}^*, K_{32}^*, K_{33}^*, ..., K_{3S_2^*}^*) \hspace{2cm} (53)$$

where, $K_{31}^* = (3, ..., \text{till } N)$ and so on K_{32}^*, K_{33}^* etc.

Thus the number of selected events characterizing third components denoted by S_3^* is the sum of numbers of indexes held by the assembly of assemblies (53) i.e.,

$$S_3^* = S_{31}^* + S_{32}^* + S_{33}^* + ... S_2^* \text{ terms.}$$

$$= \sum_{g=1}^{S_2^*} S_{3.g}^* \hspace{2cm} (54)$$

where, $S_{31}^* = $ number of indexes held by the assembly K_{31}^* and so on S_{32}^*, S_{33}^* etc.

Using the summation method we get (54) as

$$S_3^* = \sum_{k_1^* \in K_1^*} \sum_{k_2^* \in K_2^*} \sum_{k_3^* \in K_3^*} C \quad \text{————————} \quad (55)$$

Proceeding this way we get the number of selected events characterizing V^{th} components denoted by S_V^* is

$$S_V^* = S_{V1}^* + S_{V2}^* + S_{V3}^* + \ldots\ldots\ldots S_{V-1}^* \text{ terms.}$$

$$= \sum_{g=1}^{S_{V-1}^*} S_{V.g}^* \quad \text{————————} \quad (56)$$

Using the summation method we get (56) as

$$S_V^* = \sum_{k_1^* \in K_1^*} \sum_{k_2^* \in K_2^*} \sum_{k_3^* \in K_3^*} \cdots \cdots \sum_{k_V^* \in K_V^*} C \quad \text{————} \quad (57)$$

As the selected events characterizing V^{th} components are of one permutation so the equations (56) and (57) give the number of permutations of N components that not all different taken V at a time.

Example 15: How many permutations can be made by the 12-tuples (2, 2, 2, 3, 3, 3, 5, 5, 7, 7, 11, 11) taken 3 at a time.

Solution: We have given N = 12, V = 3, $N_1 = 3$, $N_2 = 3$, $N_3 = 2$, $N_4 = 2$ and $N_5 = 2$.

Now in the interval

$1 \leq k_1^* \leq$ till N

we get the assembly of assemblies of selected indexes as

$K_1^* = (K_{11}^*)$ where, $K_{11}^* = (1, 4, 7, 9, 11)$

So, $S_1^* = 5$

Again in the intervals

$1 \leq k_2^* \leq$ till $k_1^* - 1$

$k_1^* + 1 \leq k_2^* \leq$ till N

we get the assembly of assemblies of selected indexes as

$$K_2^* = (K_{21}^*, K_{22}^*, K_{23}^*, K_{24}^*, K_{25}^*) \quad \text{where, } K_{21}^* = (2, 4, 7, 9, 11)$$
$$K_{22}^* = (1, 5, 7, 9, 11)$$
$$K_{23}^* = (1, 4, 8, 9, 11)$$
$$K_{24}^* = (1, 4, 7, 10, 11)$$
$$K_{25}^* = (1, 4, 7, 9, 12)$$

So, $S_2^* = \sum_{g=1}^{5} S_{2g}^*$

$$= S_{21}^* + S_{22}^* + S_{23}^* + S_{24}^* + S_{25}^*$$

$$= 5 + 5 + 5 + 5 + 5$$
$$= 25$$

Again in the intervals

$1 \leq k_3^* \leq$ till $k_1^* - 1$ [The condition changeable and

$k_1^* + 1 \leq k_3^* \leq$ till $k_2^* - 1$ hence the intervals changeable]

$k_2^* + 1 \leq k_3^* \leq$ till N

we get the assembly of assemblies of selected indexes as

$K_3^* = (K_{31}^*, K_{32}^*, K_{33}^*, K_{34}^*, K_{35}^*, K_{36}^*, K_{37}^*, K_{38}^*, K_{39}^*, K_{3.10}^*, K_{3.11}^*,$
$K_{3.12}^*, K_{3.13}^*, K_{3.14}^*, K_{3.15}^*, K_{3.16}^*, K_{3.17}^*, K_{3.18}^*, K_{3.19}^*, K_{3.20}^*, K_{3.21}^*, K_{3.22}^*,$
$K_{3.23}^*, K_{3.24}^*, K_{3..25}^*)$

where, $K_{31}^* = (3, 4, 7, 9, 11)$; $S_{31}^* = 5$

$K_{32}^* = (2, 5, 7, 9, 11)$; $S_{32}^* = 5$

$K_{33}^* = (2, 4, 8, 9, 11)$; $S_{33}^* = 5$

$K_{34}^* = (2, 4, 7, 10, 11)$; $S_{34}^* = 5$

$K_{35}^* = (2, 4, 7, 9, 12)$; $S_{35}^* = 5$

$K_{36}^* = (2, 5, 7, 9, 11)$; $S_{36}^* = 5$

$K_{37}^* = (1, 6, 7, 9, 11)$; $S_{37}^* = 5$

$K_{38}^* = (1, 5, 8, 9, 11)$; $S_{38}^* = 5$

$K_{39}^* = (1, 5, 7, 10, 11)$; $S_{39}^* = 5$

$K_{3.10}^* = (1, 5, 7, 9, 12)$; $S_{3.10}^* = 5$

$K_{3.11}^* = (2, 4, 8, 9, 11)$; $S_{3.11}^* = 5$

$K_{3.12}^* = (1, 5, 8, 9, 11)$; $S_{3.12}^* = 5$

$K_{3.13}^* = (1, 4, 9, 11)$; $S_{3.13}^* = 4$

$K_{3.14}^* = (1, 4, 8, 10, 11)$; $S_{3.14}^* = 5$

$K_{3.15}^* = (1, 4, 8, 9, 12)$; $S_{3.15}^* = 5$

$K_{3.16}^* = (2, 4, 7, 10, 11)$; $S_{3.16}^* = 5$

$K_{3.17}^* = (1, 5, 7, 10, 11)$; $S_{3.17}^* = 5$

$K_{3.18}^* = (1, 4, 8, 10, 11)$; $S_{3.18}^* = 5$

$K_{3.19}^* = (1, 4, 7, 11)$; $S_{3.19}^* = 4$

$K_{3.20}^* = (1, 4, 7, 10, 12)$; $S_{3.20}^* = 5$

$K_{3.21}^* = (2, 4, 7, 9, 12)$; $S_{3.21}^* = 5$

$K_{3.22}^* = (1, 5, 7, 9, 12)$; $S_{3.22}^* = 5$

$K_{3.23}^* = (1, 4, 8, 9, 12)$; $S_{3.23}^* = 5$

$K_{3.24}^* = (1, 4, 7, 10, 12)$; $S_{3.24}^* = 5$

$K_{3.25}^* = (1, 4, 7, 9)$; $S_{3.25}^* = 4$

So, $S_3^* = \sum_{g=1}^{25} S_{3g}^*$

$= S_{31}^* + S_{32}^* + S_{33}^* + S_{34}^* + S_{35}^* + S_{36}^* + S_{37}^* + S_{38}^* + S_{39}^* + S_{3.10}^* +$
$S_{3.11}^* + S_{3.12}^* + S_{3.13}^* + S_{3.14}^* + S_{3.15}^* + S_{3.16}^* + S_{3.17}^* + S_{3.18}^* + S_{3.19}^* + S_{3.20}^* +$
$S_{3.21}^* + S_{3.22}^* + S_{3.23}^* + S_{3.24}^* + S_{3..25}^*$

$= 5 + 5 + 5 + 5 + 5 + 5 + 5 + 5 + 5 + 5 + 5 + 5 + 4 + 5 + 5 + 5 + 5$
$+ 5 + 4 + 5 + 5 + 5 + 5 + 5 + 4$

$= 122.$

Thus the desired number of permutations is

$P^* \binom{12}{3} = 122.$

Using the summation method we get the desired number of

$P^* \binom{12}{3} = \sum_{k_1^* \in K_1^*} \sum_{k_2^* \in K_2^*} \sum_{k_3^* \in K_3^*} C$

$= \sum_{k_2^* \in K_{21}^*} \sum_{k_3^* \in K_3^*} C + \sum_{k_2^* \in K_{22}^*} \sum_{k_3^* \in K_3^*} C + \sum_{k_2^* \in K_{23}^*} \sum_{k_3^* \in K_3^*} C$

$\quad + \sum_{k_2^* \in K_{24}^*} \sum_{k_3^* \in K_3^*} C + \sum_{k_2^* \in K_{25}^*} \sum_{k_3^* \in K_3^*} C$

$= (\sum_{k_3^* \in K_{31}^*} C + \sum_{k_3^* \in K_{32}^*} C + \sum_{k_3^* \in K_{33}^*} C + \sum_{k_3^* \in K_{34}^*} C + \sum_{k_3^* \in K_{35}^*} C)$

$+ (\sum_{k_3^* \in K_{36}^*} C + \sum_{k_3^* \in K_{37}^*} C + \sum_{k_3^* \in K_{38}^*} C + \sum_{k_3^* \in K_{39}^*} C + \sum_{k_3^* \in K_{3.10}^*} C)$

$+ (\sum_{k_3^* \in K_{3.11}^*} C + \sum_{k_3^* \in K_{3.12}^*} C + \sum_{k_3^* \in K_{3.13}^*} C + \sum_{k_3^* \in K_{3.14}^*} C + \sum_{k_3^* \in K_{3.15}^*} C)$

$+ (\sum_{k_3^* \in K_{3.16}^*} C + \sum_{k_3^* \in K_{3.17}^*} C + \sum_{k_3^* \in K_{3.18}^*} C + \sum_{k_3^* \in K_{3.19}^*} C + \sum_{k_3^* \in K_{3.20}^*} C)$

$+ (\sum_{k_3^* \in K_{3.21}^*} C + \sum_{k_3^* \in K_{3.22}^*} C + \sum_{k_3^* \in K_{3.23}^*} C + \sum_{k_3^* \in K_{3.24}^*} C + \sum_{k_3^* \in K_{3.25}^*} C)$

$= 5 + 5 + 5 + 5 + 5 + 5 + 5 + 5 + 5 + 5 + 5 + 4 + 5 + 5 + 5$
$\quad + 5 + 5 + 4 + 5 + 5 + 5 + 5 + 5 + 4$

$= 122.$

15. Identified Selected Permutation Theorem

Theorem 9: The number of permutations of N components that not all different and there are N_1 components alike of one kind, N_2 components alike of another kind, N_3 components alike of a third kind and so on for a hth kind taken V at a time, whose first v components are identified denoted by $P^* \binom{N}{V/^v A}$ is

$P^* \binom{N}{V/^v A} = \sum_{g=1}^{S_{V-1/v}^*} S_{V/v.g}^*$ \underline{\hspace{3cm}} (58)

where, $V = v+1, v+2, v+3, \ldots\ldots, N$

$N = N_1 + N_2 + N_3 + \ldots\ldots + N_h$

$S_v^* = 1.$

Using the summation method we get (58) as

$$P^* \left(\frac{N}{V/^v A} \right) = \Sigma_{k_{v+1}^* \in K_{v+1}^*} \Sigma_{k_{v+2}^* \in K_{v+2}^*} \Sigma_{k_{v+3}^* \in K_{v+3}^*} \cdots \cdots \Sigma_{k_v^* \in K_v^*} C \quad \text{------ (59)}$$

where C is a constant quantity taking unit value.

Proof: Suppose we have N components that not all different. Let there are N_1 components alike of one kind, N_2 components alike of another kind, N_3 components alike of a third kind and so on N_h components alike of a hth kind, then we have k_{v+1} the index that indicates a component of the O. P. A. taken by $(v+1)^{th}$ place of a permutation, as

$1 \leq k_{v+1} \leq k_1 - 1$

$k_1 + 1 \leq k_{v+1} \leq k_2 - 1$

$k_2 + 1 \leq k_{v+1} \leq k_3 - 1$ [See (21)]

$\vdots$

$k_v + 1 \leq k_{v+1} \leq N$

subject to the condition $1 < k_1 < k_2 \ldots\ldots < k_v < N$. As the condition may be changed so the intervals may be changed.

Now k_{v+1}^* the index that indicates the components of the O. P. A. which is not alike of previous components indicated by previous indexes of the intervals (21) as

$1 \leq k_{v+1}^* \leq$ till $k_1^* - 1$

$k_1^* + 1 \leq k_{v+1}^* \leq$ till $k_2^* - 1$

$k_2^* + 1 \leq k_{v+1}^* \leq$ till $k_3^* - 1$ (60)

$\vdots$

$k_v^* + 1 \leq k_{v+1}^* \leq$ till N

subject to the condition $1 < k_1^* < k_2^* < \ldots\ldots < k_v^* < N$. As the condition may be changed so the intervals may be changed. As $k_1^*, k_2^*, k_3^*, \ldots\ldots, k_v^*$ identified then suppose

$k_1^* = 1, k_2^* = 2, k_3^* = 3, \ldots\ldots, k_v^* = v$

Then the selected indexes for k_{v+1}^* are

$v+1, \ldots\ldots,$ till N

The dots after comma takes only selected indexes. Thus the assembly of assemblies of selected indexes is

$$K_{v+1}^* = (K_{(v+1).1}^*) \qquad\qquad (61)$$

$$\text{where, } K_{(v+1).1}^* = (\ v+1, \ \ldots, \text{ till N })$$

Now the number of selected events characterizing $(v+1)^{th}$ components whose first v components are identified denoted by $S_{v+1/v}^*$ is the number of indexes held by the assembly of assemblies (61) i.e.,

$$S_{v+1/v}^* = S_{v+1/v.1}^* = \sum_{g=1}^{S_{v/v}^*} S_{v+1/v.g}^* \qquad\qquad (62)$$

$$\text{where } S_{v/v}^* = 1$$

and $S_{v+1/v.1}^* = $ number of indexes held by the assembly K_{v+1}^*.

Using the summation method we get (62) as

$$S_{v+1/v}^* = \sum_{k_{v+1}^* \in K_{v+1}^*} C \qquad\qquad (63)$$

where C is a constant quantity taking unit value.

The change in $k_1^*, k_2^*, k_3^*, \ldots\ldots, k_v^*$ give you change in the number of particular intervals for (60) but does not effect in the number of indexes held in (60).

Again we have k_{v+2} the index that indicates a component of the O. P. A. taken by $(v+2)^{th}$ place of a permutation as

$1 \le k_{v+2} \le k_1-1$

$k_1+1 \le k_{v+2} \le k_2-1$

$k_2+1 \le k_{v+2} \le k_3-1$ $\qquad\qquad$ [See (24)]

$\vdots$

$k_v+1 \le k_{v+2} \le k_{v+1}-1$

$k_{v+1}+1 \le k_{v+2} \le N$

Subject to the condition $1 < k_1 < k_2 \ldots\ldots < k_{v+1} < N$. If the condition changes then the intervals will be changed.

Now k_{v+2}^* the index that indicates the component of the O. P. A. which is not alike of previous components indicated by previous indexes contained in the intervals (24) as

$1 \le k_{v+2}^* \le \text{ till } k_1^* -1$

$k_1^* +1 \le k_{v+2}^* \le \text{ till } k_2^* -1$

$k_2^* +1 \le k_{v+2}^* \le \text{ till } k_3^* -1$ $\qquad\qquad (64)$

$\vdots$

$k_v^* + 1 \leq k_{v+2}^* \leq$ till $k_{v+1}^* - 1$

$k_{v+1}^* + 1 \leq k_{v+2}^* \leq$ till N

Subject to the condition $1 < k_1^* < k_2^* < \ldots\ldots < k_{v+1}^* < N$. If the condition changes then the intervals will be changed.

As $k_1^*, k_2^*, k_3^*, \ldots\ldots, k_v^*$ are identified then suppose

$k_1^* = 1, k_2^* = 2, k_3^* = 3, \ldots\ldots, k_v^* = v$

Thus the selected indexes for k_{v+2}^* are i.e., when

$k_{v+1}^* = v + 1$ then k_{v+2}^* takes $v+2, \ldots\ldots$, till N

$k_{v+1}^* =$ till N then k_{v+2}^* takes $v+1, \ldots.$, till $(N_1+N_2+N_3+ \ldots.. +N_{h-1})$

The dots after comma takes only selected indexes. Now the assembly of assemblies of selected indexes is

$$K_{v+2}^* = (K_{(v+2).1}^*, K_{(v+2).2}^*, K_{(v+2).3}^*, \ldots\ldots, K_{(v+2)S_{v+1/v}^*}^*) \;\underline{\quad\quad}\; (65)$$

where, $K_{(v+2).1}^* = (v+2, \ldots.., $ till N $)$

and so on $K_{(v+2)S_{v+1/v}^*}^* = (v+1, \ldots\ldots, $ till $(N_1+N_2+N_3+ \ldots\ldots +N_{h-1}))$.

Thus the number of selected events characterizing $(v+2)^{th}$ components whose first v components are identified, denoted by $S_{v+2/v}^*$ is the sum of numbers of indexes held by the assembly of assemblies (65) i.e.,

$$S_{v+2/v}^* = S_{v+2/v.1}^* + S_{v+2/v.2}^* + S_{v+2/v.3}^* + \ldots\ldots S_{v+1/v}^* \text{ terms}$$

$$= \sum_{g=1}^{S_{v+1/v}^*} S_{v+2/v.g}^* \;\underline{\quad\quad\quad}\; (66)$$

where, $S_{v+2/v.1}^* =$ number of indexes held by the assembly $K_{(v+2).1}^*$ and so on for $S_{v+2/v.2}^*, S_{v+2/v.3}^*$ etc.

Using the summation method we get (66) as

$$S_{v+2/v}^* = \sum_{k_{v+1}^* \in K_{v+1}^*} \sum_{k_{v+2}^* \in K_{v+2}^*} C \;\underline{\quad\quad\quad}\; (67)$$

The change in $k_1^*, k_2^*, k_3^*, \ldots\ldots, k_v^*$ give you change in the number of particular intervals for (64) but does not effect in the number of indexes held in (64).

Now we have k_{v+3} the index that indicates a component of the O. P. A. taken by $(v+3)^{th}$ place of the permutation as

$1 \leq k_{v+3} \leq k_1 - 1$

$k_1 + 1 \leq k_{v+3} \leq k_2 - 1$

$k_2 + 1 \leq k_{v+3} \leq k_3 - 1$

$\vdots$ [See (27)]

$$k_v + 1 \leq k_{v+3} \leq k_{v+1} - 1$$
$$k_{v+1} + 1 \leq k_{v+3} \leq k_{v+2} - 1$$
$$k_{v+2} + 1 \leq k_{v+3} \leq N$$

subject to the condition $1 < k_1 < k_2 \ldots\ldots < k_{v+2} < N$. If the condition changes then the intervals will be changed.

Again k_{v+3}^* the index that indicates the component of the O. P. A. which is not alike of previous components indicated by previous indexes contained in the intervals (27) as

$$1 \leq k_{v+3}^* \leq \text{ till } k_1^* - 1$$
$$k_1^* + 1 \leq k_{v+3}^* \leq \text{ till } k_2^* - 1$$
$$k_2^* + 1 \leq k_{v+3}^* \leq \text{ till } k_3^* - 1$$
$$\vdots$$

$$k_v^* + 1 \leq k_{v+3}^* \leq \text{ till } k_{v+1}^* - 1$$
$$k_{v+1}^* + 1 \leq k_{v+3}^* \leq \text{ till } k_{v+2}^* - 1$$
$$k_{v+2}^* + 1 \leq k_{v+3}^* \leq \text{ till } N$$

$$\text{(68)}$$

subject to the condition $1 < k_1^* < k_2^* < \ldots\ldots < k_{v+2}^* < N$. If the condition change then the intervals will be changed. As $k_1^*, k_2^*, k*_3, \ldots\ldots, k_v^*$ are identified then suppose

$$k_1^* = 1, \quad k_2^* = 2, \quad k_3^* = 3, \ldots\ldots, k_v^* = v$$

Then the selected indexes for k_{v+3}^* are i.e., when

$k_{v+1}^* = v+1$ and $k_{v+2}^* = v+2$ then k_{v+3}^* takes $v+3, \ldots\ldots,$ till N
$$\vdots$$

$k_{v+1}^* = v+1$ and $k_{v+2}^* =$ till N then k_{v+3}^* takes
$$v+2, \ldots\ldots, \text{ till } (N_1 + N_2 + N_3 + \ldots\ldots + N_{h-1})$$
$$\vdots$$

$k_{v+1}^* =$ till N and $k_{v+2}^* = v+1$ then k_{v+3}^* takes
$$v+2, \ldots\ldots, \text{ till } (N_1 + N_2 + N_3 + \ldots\ldots + N_{h-1})$$
$$\vdots$$

$k_{v+1}^* =$ till N and $k_{v+2}^* =$ till $(N_1 + N_2 + N_3 + \ldots\ldots + N_{h-1})$ then k_{v+3}^* takes
$$v+1, \ldots\ldots, \text{ till } (N_1 + N_2 + N_3 + \ldots\ldots + N_{h-2})$$

The dots after comma takes only selected indexes. Now the assembly of assemblies of selected indexes is

$$K_{v+3}^* = (K_{(v+3).1}^*, \ K_{(v+3).2}^*, \ K_{(v+3).3}^*, \cdots \cdots, \ K_{(v+3)S_{v+2/v}^*}^*) \quad \text{(69)}$$

where, $K_{(v+3).1}^* = (v+3, \ldots\ldots, \text{ till } N)$

and so on for $K^*_{(v+3).2}$, $K^*_{(v+3).3}$ etc.

Thus the number of selected events characterizing $(v+3)^{th}$ components whose first v components are identified, denoted by $S^*_{v+3/_v}$ is the sum of numbers of indexes held by the assembly of assemblies (69) i.e.,

$$S^*_{v+3/_v} = S^*_{v+3/_{v.1}} + S^*_{v+3/_{v.2}} + S^*_{v+3/_{v.3}} + \ldots\ldots S^*_{v+2/_v} \text{ terms}$$

$$= \sum_{g=1}^{S^*_{v+2/v}} S^*_{v+3/_{v.g}} \quad\quad\quad\quad (70)$$

where, $S^*_{v+3/_{v.1}}$ = number of indexes held by the assembly $K^*_{(v+3).1}$

and so on for $S^*_{v+3/_{v.2}}$, $S^*_{v+3/_{v.3}}$ etc.

Using the summation method we get (70) as

$$S^*_{v+3/_v} = \sum_{k^*_{v+1} \in K^*_{v+1}} \sum_{k^*_{v+2} \in K^*_{v+2}} \sum_{k^*_{v+3} \in K^*_{v+3}} C \quad\quad (71)$$

Here the change in k^*_1, k^*_2, k^*_3, $\ldots\ldots$, k^*_v give you change in the number of particular intervals for (68) but does not effect in the number of indexes held in (68).

Proceeding this way we get the number of selected events characterizing v^{th} components whose first v components are identified, denoted by $S^*_{V/_v}$ is

$$S^*_{V/_v} = S^*_{V/_{v.1}} + S^*_{V/_{v.2}} + S^*_{V/_{v.3}} + \ldots\ldots S^*_{V-1/_v} \text{ terms}$$

$$= \sum_{g=1}^{S^*_{V-1/v}} S^*_{V/_{v.g}} \quad\quad\quad\quad (72)$$

Using the summation method we get (72) as

$$S^*_{V/_v} = \sum_{k^*_{v+1} \in K^*_{v+1}} \sum_{k^*_{v+2} \in K^*_{v+2}} \sum_{k^*_{v+3} \in K^*_{v+3}} \cdots \cdots \sum_{k^*_V \in K^*_V} C \quad (73)$$

As the selected events characterizing V^{th} components are of one permutation so the equations (72) and (73) give the number of permutations of N components that not all different taken V at a time whose first v components are identified.

Corollary 3: The number of permutations of N components that not all different taken V at a time, whose first v components are identified, is the same as the number of permutations of (N−v) components that all different (or not) taken (V−v) at a time i.e.,

$$P^*\begin{pmatrix} N \\ V/^vA \end{pmatrix} = P^*\begin{pmatrix} N-v \\ V-v \end{pmatrix} \quad\quad\quad\quad (74)$$

where (N−v) components chosen from N components eliminating v identified components.

Proof: We have N components that not all different. The permutations are of V components whose first v components are identified. Suppose the v identified components eliminated from N components then (N−v) components remain. Now the number of permutations of (N−v) components taken (V−v) components is $P^* \begin{pmatrix} N-v \\ V-v \end{pmatrix}$. Again v eliminated identified components put before each permutation. Then the permutations be of V components taken from N components whose first v components are identified and the number is $P^* \begin{pmatrix} N \\ V/^vA \end{pmatrix}$. But the number is the same as $P^* \begin{pmatrix} N-v \\ V-v \end{pmatrix}$. Thus $P^* \begin{pmatrix} N \\ V/^vA \end{pmatrix} = P^* \begin{pmatrix} N-v \\ V-v \end{pmatrix}$.

16. Conclusion

From this paper we get idea about permutation theorem as well as identified permutation theorem. Apart from this, we gain knowledge about the very important general permutation theorem and selected permutation theorem.

Acknowledgment

To compose this paper I have taken help from my book 'Bystematics Vol. II My Classic', Scholar's Press, 29 March 2018, ISBN: 978- 620-2-30960-8.

References

1. Bystematics Vol. I My Classic by Deapon Biswas, Scholar's Press 29 March 2018, ISBN: 978- 620-2-30664-5

2. Bystematics Vol. II My Classic by Deapon Biswas, Scholar's Press 29 March 2018, ISBN: 978- 620-2-30960-8.

3. F. Mosteller, R. E. K Rourke & G. B. Thomas Jr. Probability with Statistical Applications.

YOUR KNOWLEDGE HAS VALUE